the SCIENCE library

PLANTS

the S C I E N C E *library*

PLANTS

Peter Riley
Consultant: Steve Parker

This 2009 edition published and distributed by:

Mason Crest Publishers Inc.

370 Reed Road, Broomall, Pennsylvania 19008

(866) MCP-BOOK (toll free)

www.masoncrest.com

Library of Congress Cataloging-in-Publication data is available

Plants
ISBN 978-1-4222-1554-8

The Science Library - 10 Title Series
ISBN 978-1-4222-1546-3

Printed in the United States of America

First published in 2004 by Miles Kelly Publishing Ltd
Bardfield Centre Great Bardfield Essex CM7 4SL

Editorial Director Belinda Gallagher

Art Director Jo Brewer

Editor Jenni Rainford

Editorial Assistant Chloe Schroeter

Cover Design Simon Lee

Design Concept Debbie Meekcoms

Design Stonecastle Graphics

Consultant Steve Parker

Indexer Hilary Bird

Reprographics Stephan Davis, Ian Paulyn

Production Manager Elizabeth Brunwin

Contents

How to use this book

PLANTS is packed with information, color photos, diagrams, illustrations and features to help you learn more about science. Do you know what the oldest tree is or how bees help plants to reproduce? Did you know the rafflesia flower's petals can grow to a width of 1 meter or that there are 32 million tons of apples grown worldwide every year? Enter the fascinating world of science and learn about why things happen, where things come from and how things work. Find out how to use this book and start your journey of scientific discovery.

Main text
Each page begins with an introduction to the different subject areas.

Woody plants

THE ROOTS and shoots of most plants are made of soft materials, but some plants make their roots and shoots from a very hard material – wood. These woody plants are trees and bushes. Unlike other plants that die at the end of the growing season, wood is so strong that it allows the plant to keep on growing and become larger and larger every year. The woody stems are covered in bark to protect the tree from animals that might try to eat it. Bark also protects the tree from the bitterly cold winter weather by making a warm insulating layer over the surface.

● **Holding on to leaves**
There are two kinds of trees: evergreen, which hold on to their leaves all year round, and deciduous, which usually lose their leaves in autumn and grow new ones the following spring. An evergreen tree does lose some leaves during the year but it does not lose them all at once; it always has some leaves on its twigs to make it look green

▶▶ Read further > forests
pg29 (h22, i32)

To scale
This coast redwood is the fourth highest tree ever measured – it is almost the length of a soccer pitch

Each square = 25 m

● Check it out!
• http://www.treecouncil.org.uk

Coast redwood is about 112 m high

Length of average soccer pitch is 92 m

Some pine needles can grow to a length of 30 cm

1 2 3 4 5 6 7 8 9 10 11 12 13 14 15

To scale
Some pages show the size of an object in relation to another so that you can compare how big or small things really are.

Check it out!
Find out more by surfing the Internet.

Main image
Each topic is clearly illustrated. Some images are labeled, providing further information.

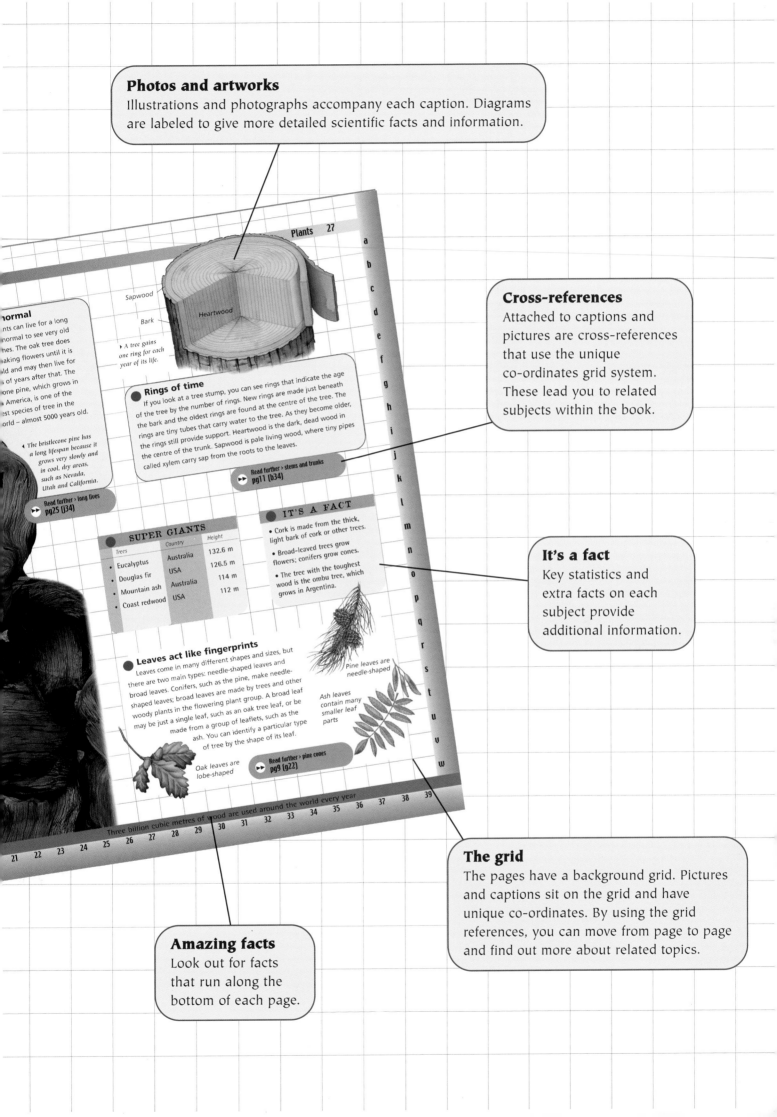

Photos and artworks
Illustrations and photographs accompany each caption. Diagrams are labeled to give more detailed scientific facts and information.

Cross-references
Attached to captions and pictures are cross-references that use the unique co-ordinates grid system. These lead you to related subjects within the book.

It's a fact
Key statistics and extra facts on each subject provide additional information.

The grid
The pages have a background grid. Pictures and captions sit on the grid and have unique co-ordinates. By using the grid references, you can move from page to page and find out more about related topics.

Amazing facts
Look out for facts that run along the bottom of each page.

Plants 27

Sapwood

Bark

Heartwood

▸ *A tree gains one ring for each year of its life.*

...normal
...nts can live for a long ...normal to see very old ...hes. The oak tree does ...aking flowers until it is ...old and may then live for ...s of years after that. The ...one pine, which grows in ... America, is one of the ...est species of tree in the ...orld – almost 5000 years old.

◂ *The bristlecone pine has a long lifespan because it grows very slowly and in cool, dry areas, such as Nevada, Utah and California.*

▸▸ Read further › long lives
pg25 (j34)

Rings of time
If you look at a tree stump, you can see rings that indicate the age of the tree by the number of rings. New rings are made just beneath the bark and the oldest rings are found at the centre of the tree. The rings are tiny tubes that carry water to the tree. As they become older, the rings still provide support. Heartwood is the dark, dead wood in the centre of the trunk. Sapwood is pale living wood, where tiny pipes called xylem carry sap from the roots to the leaves.

▸▸ Read further › stems and trunks
pg11 (b34)

SUPER GIANTS

Trees	Country	Height
• Eucalyptus	Australia	132.6 m
• Douglas fir	USA	126.5 m
• Mountain ash	Australia	114 m
• Coast redwood	USA	112 m

IT'S A FACT
• Cork is made from the thick, light bark of cork or other trees.
• Broad-leaved trees grow flowers; conifers grow cones.
• The tree with the toughest wood is the ombu tree, which grows in Argentina.

Pine leaves are needle-shaped

Ash leaves contain many smaller leaf parts

Leaves act like fingerprints
Leaves come in many different shapes and sizes, but there are two main types: needle-shaped leaves and broad leaves. Conifers, such as the pine, make needle-shaped leaves; broad leaves are made by trees and other woody plants in the flowering plant group. A broad leaf may be just a single leaf, such as an oak tree leaf, or be made from a group of leaflets, such as the ash. You can identify a particular type of tree by the shape of its leaf.

Oak leaves are lobe-shaped

▸▸ Read further › pine cones
pg9 (g22)

Three billion cubic metres of wood are used around the world every year

21 22 23 24 25 26 27 28 29 30 31 32 33 34 35 36 37 38 39

a b c d e f g h i j k l m n o p q r s t u v w

The plant kingdom

THERE ARE millions of different kinds of living things on the Earth, so scientists group them to make it easier to study. One of the largest groups is the plant kingdom, with more than 400,000 different kinds of plants. Some plants are so small that you need a microscope to see them clearly, while the tallest trees can reach over 100 m in height (*see pg26 [r8]*). Plants also vary in how long they live (*see pg27 [j23]*). Some live for just a few hours, while others can live for thousands of years.

▶ *There are many different kinds, or groups, of plants and other plant-based living things. The main groups are shown here.*

Flowering plants

Broadleaved trees and bushes, flowers and herbs

Gingkos

Conifers

Cycads

Ferns

Club mosses

Horsetails

Mosses

Liverworts

Fungi

Lichens

Algae and seaweeds

Microscopic plants

● Plants with flowers

More than 250,000 types of plants grow flowers. They are known as the flowering plant group. Monocotyledons and dicotyledons are the two types of flowering plants (*see pg16 [d2]*). Many flowers have brightly colored petals (*see pg18 [p9]*) to attract insects to pollinate them. Other types use methods such as self-pollination (*see pg 19 [i27]*), or wind pollination (*see pg 19 [u34]*). The flowers make seeds that grow into new plants.

● **Check it out!**
• http://www.kew.org

▶▶ **Read further › monocotyledons and dicotyledons / insect pollination** pg16 [d2; j14]; pg18 [k10]; pg19 [b29; d2]

The Venus flytrap plant feeds on insects

IT'S A FACT

• The plant with the largest flower is the rafflesia, which grows up to 1 m across.

• The welwitschia plant of the southern African scrub has two leaves, each many meters long, which last for hundreds of years.

Plants with cones

Conifers are trees that have scaly or needle-like leaves *(see pg27 [q36])* that they keep all year round. They do not have flowers but grow two types of cone – one makes a yellow dust called pollen *(see pg19 [u34])*, the other makes seeds when it receives the pollen. The cones that contain the seeds are made from a woody material that has slits in it. When the seeds are ready to leave, the slits open and the seeds fall out.

▲ *Cones grow near the tips of conifer branches.*

Read further › evergreens / conifers
pg26 (l12); pg27 (q27)

BIGGEST PLANTS

Type	Plant	Length/weight
• Tallest grass	Bamboo	25 m
• Tallest cactus	Saguaro	18 m
• Biggest fern	Norfolk Island tree fern	20 m
• Biggest seed	Coco-de-mer palm	30 kg
• Longest leaf	Raffia palm	20 m

Shady plants

Mosses, ferns and liverworts are plants that grow in damp, shady places. They do not make seeds or have vessels to carry water. Mosses grow stalks with swollen tips. The tips make spores, which float away in the air. When a spore lands in damp soil it grows into a new moss plant. Ferns' spores form in button-like swellings on the underside of their fronds (that resemble feathery leaves).

Read further › spores
pg32 (k2)

▲ *Ferns on a forest floor grow close together, making thick vegetation.*

Plants in the sea

Seaweeds grow mainly on rocky shores or underwater close to the coast. They hold on to the rocks to prevent being washed away. The part that holds the seaweed to a rock is called a holdfast, and looks like a root. It grips the rock tightly like a sucker. Seaweeds have tough, leathery fronds (leaves) to stand up to the pounding of the waves. Their bodies are flexible so that they can move with the water currents without breaking.

▸ *Seaweeds such as kelp belong to the plant group called algae.*

Read further › roots / underwater plants / lichens
pg11 (b22); pg15 (b34); pg32 [d2]

Plant parts

THERE ARE four major parts to a flowering plant: the roots; the stem; the leaves; and the flower. Each plays a vital role in keeping the plant alive. The roots grow into the ground, ensuring the plant does not blow away; the stem contains small tubes that carry water to all parts of the plant; the leaves use the energy from sunlight to make food (*see pg14 [j14]*) and the flower makes seeds from which new plants grow.

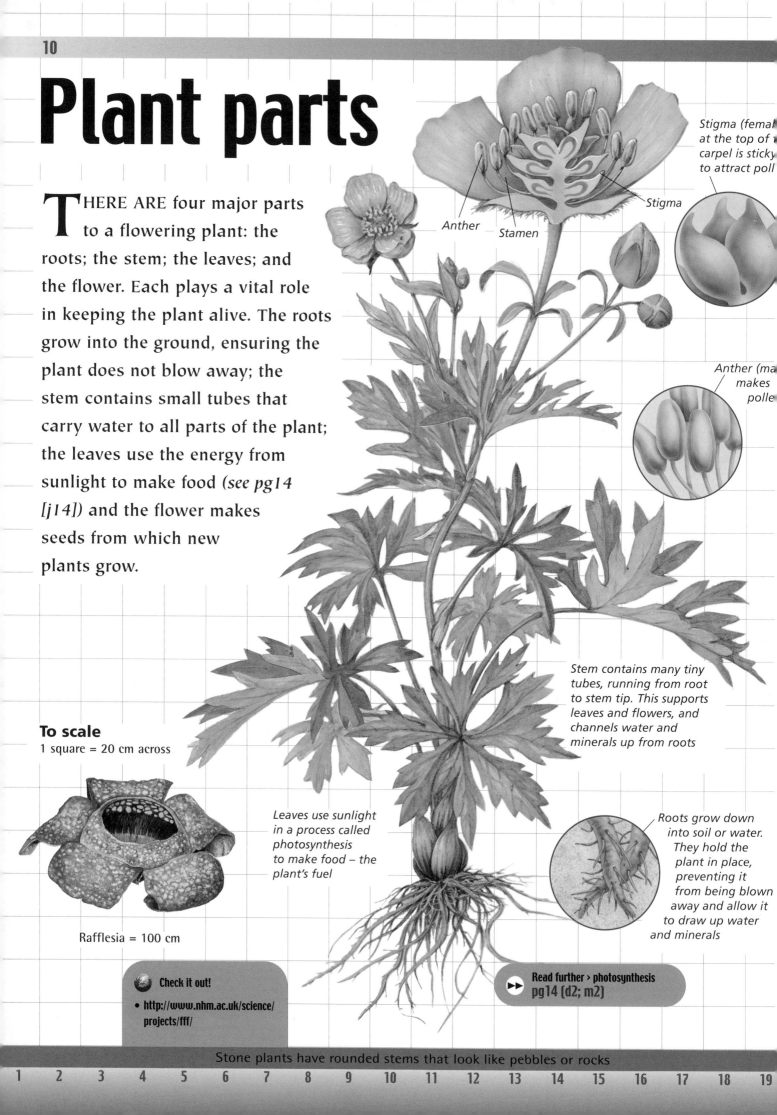

Stigma (femal at the top of carpel is sticky to attract poll

Stigma

Anther   Stamen

Anther (ma makes polle

Stem contains many tiny tubes, running from root to stem tip. This supports leaves and flowers, and channels water and minerals up from roots

To scale
1 square = 20 cm across

Rafflesia = 100 cm

Leaves use sunlight in a process called photosynthesis to make food – the plant's fuel

Roots grow down into soil or water. They hold the plant in place, preventing it from being blown away and allow it to draw up water and minerals

Check it out!
• http://www.nhm.ac.uk/science/
projects/fff/

▶▶ **Read further › photosynthesis**
pg14 [d2; m2]

Stone plants have rounded stems that look like pebbles or rocks

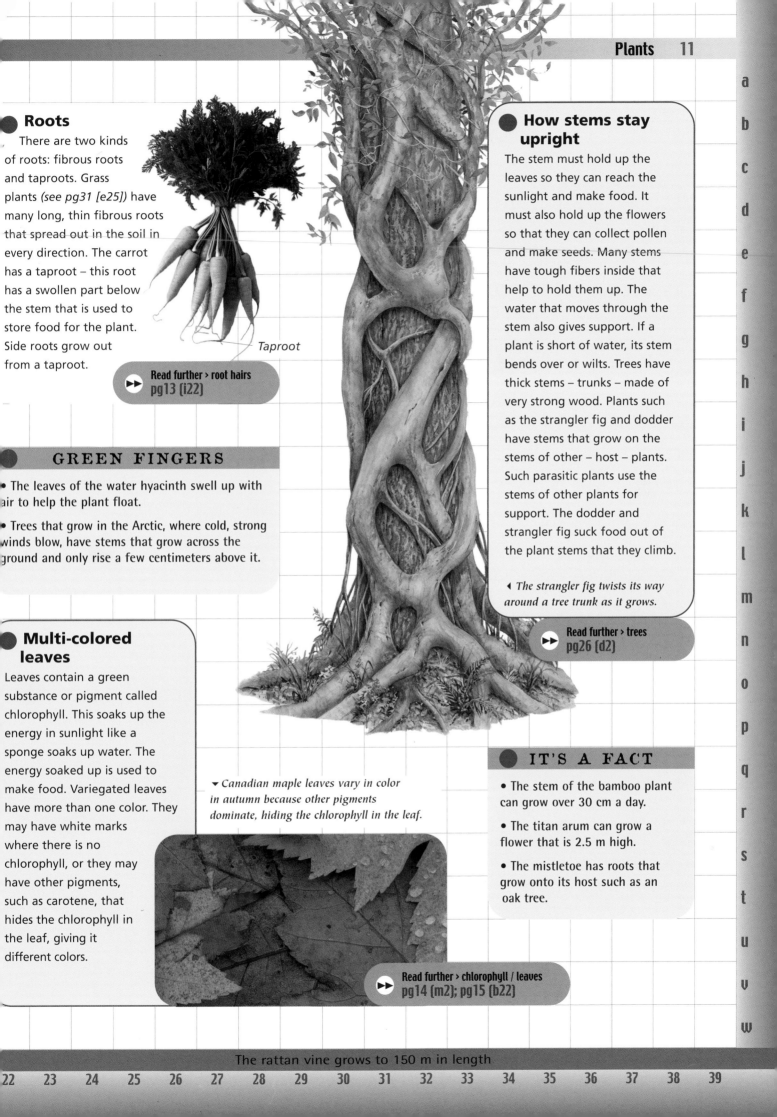

Roots

There are two kinds of roots: fibrous roots and taproots. Grass plants *(see pg31 [e25])* have many long, thin fibrous roots that spread out in the soil in every direction. The carrot has a taproot – this root has a swollen part below the stem that is used to store food for the plant. Side roots grow out from a taproot.

Taproot

Read further > root hairs
pg13 [i22]

GREEN FINGERS

- The leaves of the water hyacinth swell up with air to help the plant float.

- Trees that grow in the Arctic, where cold, strong winds blow, have stems that grow across the ground and only rise a few centimeters above it.

Multi-colored leaves

Leaves contain a green substance or pigment called chlorophyll. This soaks up the energy in sunlight like a sponge soaks up water. The energy soaked up is used to make food. Variegated leaves have more than one color. They may have white marks where there is no chlorophyll, or they may have other pigments, such as carotene, that hides the chlorophyll in the leaf, giving it different colors.

▼ *Canadian maple leaves vary in color in autumn because other pigments dominate, hiding the chlorophyll in the leaf.*

Read further > chlorophyll / leaves
pg14 [m2]; pg15 [b22]

How stems stay upright

The stem must hold up the leaves so they can reach the sunlight and make food. It must also hold up the flowers so that they can collect pollen and make seeds. Many stems have tough fibers inside that help to hold them up. The water that moves through the stem also gives support. If a plant is short of water, its stem bends over or wilts. Trees have thick stems – trunks – made of very strong wood. Plants such as the strangler fig and dodder have stems that grow on the stems of other – host – plants. Such parasitic plants use the stems of other plants for support. The dodder and strangler fig suck food out of the plant stems that they climb.

◄ *The strangler fig twists its way around a tree trunk as it grows.*

Read further > trees
pg26 [d2]

IT'S A FACT

- The stem of the bamboo plant can grow over 30 cm a day.

- The titan arum can grow a flower that is 2.5 m high.

- The mistletoe has roots that grow onto its host such as an oak tree.

The rattan vine grows to 150 m in length

22 23 24 25 26 27 28 29 30 31 32 33 34 35 36 37 38 39

Thirsty work

A LL LIVING things need water to stay alive. Nearly three-quarters of a plant is made up of water, and if it loses too much it simply dies. Plants need water to make food, too, so if a plant cannot get enough water it is also in danger of starving to death. The water inside many plants helps to hold up the stem and the leaves so they can reach the sunlight to make food. If they are short of water, the stem collapses and the leaves cannot get the sunlight the plant needs to make food.

● Moving water without energy

Plants do not use any energy in sucking up water or pumping it to their leaves. Water passes into the roots of a plant from the soil. Inside the root, the water is pulled through the plant to the leaves in tiny tubes or xylem *(see pg 13 [n35])*; dissolved food from the leaves to other parts is carried by phloem *(see pg13 [o34])*. Water that evaporates (transpiration) through holes in the leaves is replaced with more water as it moves up the stem.

▶▶ **Read further › roots / stems**
pg11 (b22, b34)

● IT'S A FACT

• The tip of a root is covered with a cap of slimy cells. The root cap protects the root tips from being worn away as they grow through the ground.

• Air holes in a leaf open during the day and close at night.

Water escapes from leaves as water vapor

Water flows up inside stem through tube-like channels called xylem

▸ *The path of water in a rose plant. Water input, flow and loss in a plant is called translocation.*

Roots spread out in soil to collect water

Water enters plant through its roots

Bugs, grubs and caterpillars feed on leaves

Thicker tubes (xylem) carry water to leaf

How water escapes

Inside the leaf are air spaces. When water reaches the air holes most of it evaporates and forms a gas called water vapor. When the water vapor evaporates it passes through holes called stomata on the underside of the leaf. More water is drawn up the plant to take its place. If the air outside the leaf is hot and dry, water vapor escapes quickly so the plant needs plenty of water to keep up the supply.

Leaf pores (stomata) where water escapes

Thinner tubes (phloem) carry sap away from leaf

Air spaces in leaf

Microscopic view of holes in a leaf.

Hairy roots

Near the tip of each root are tiny, delicate hairs. They grow out a short way into the soil and collect water. The water moves along the hair and into the root where it enters a water tube. As the tip of the root grows through the soil, the hairs further back die and new hairs grow at the root tip.

GREEN FINGERS

• Leaves are covered in a layer of wax to prevent too much water from being lost or gained.

• The leaves of underwater plants take in water through their surfaces.

Xylem vessels (tubes) carry water

Growing layer

Phloem vessels (tubes) carry sap

Water carriers

If you break open a celery stalk, you will see fibers sticking out of the end. These fibers are made from groups of tubes *(see pg15 [f30])* that carry water through the plant. Groups of tubes like these are called vessels. Inside the vessels are coils of cellulose. These help to keep the tubes open so that water can always pass along them.

Strong fibers

Base of stem

Main root

Secondary root

Branching root tip

▶▶ **Read further › water storage**
pg31 (b30)

◀ *Roots grow thickly to take in large amounts of water.*

▸ *Stems have many tubes or vessels for carrying food and water.*

▶▶ **Read further › fibrous and taproots**
pg11 (b22)

Check it out!
• http://www.ars.usda.gov/is/kids

The leaves of the Amazonian giant water lily are so large and strong you can stand on them without sinking

a b c d e f g h i j k l m n o p q r s t u v w

Making food

UNLIKE ANIMALS, plants are able to make their own food. Plants get energy to make food from sunlight. Certain chemicals are needed to make food and the plants get them from the water in the ground and the air around them. When food is made, some of the energy from the sunlight is stored in it. The plants use this energy to keep them alive and growing. The whole process is called photosynthesis, which in Greek means 'building with light.'

IT'S A FACT

• During photosynthesis the oxygen in the air, on which we depend for life, is continually replenished by plants.

• Photosynthesis occurs in leaves in two special kinds of cell: palisade and spongy cells.

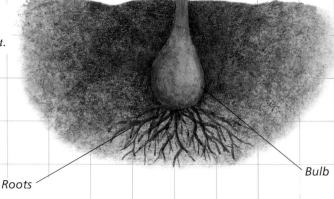

Sun's rays carry light energy to leaf

Carbon dioxide is taken from air into leaves

Oxygen is given off from leaves

Stem

Bulb

Roots

Using sunlight

Most plant leaves contain a green chemical called chlorophyll. This traps some of the energy in sunlight. During photosynthesis, plants spread out their leaves to make a large area of the chlorophyll in which the sunlight shines. Once the energy is trapped, it is used to split up water into two chemicals called hydrogen and oxygen. The plant then uses more energy to join the hydrogen with carbon dioxide in the air to make food substances called carbohydrates, mainly sugars and starches. It gives off the oxygen into the air.

▶▶ Read further › chlorophyll
pg11 (n22)

▶ *Photosynthesis occuring in a tulip plant – changing simple chemicals into food using sunlight.*

GREEN FINGERS

• Huge numbers of algae in the seas make most of the oxygen we breathe, even for those people who live far from the coast.

• Carbohydrates are foods that contain the chemicals carbon, hydrogen and oxygen.

• Two important minerals that plants take from the soil are made from chemicals called nitrogen and phosphorus. Plants use them to make protein foods.

Check it out!

• http://www.alienexplorer.com/ecology/topic3.htm

Cells in a leaf

Plants are made from tiny blocks of living matter called cells. There are different kinds of cell. The surface of the leaf is made from flat cells, which are transparent to let the sunlight shine through. Inside the leaf are cells that contain the chlorophyll. It is these cells that trap sunlight and make food. On the underside of the leaf are pairs of banana-shaped cells, which form the holes through which water vapor and carbon dioxide pass.

Read further > water vapour
pg13 (b34)

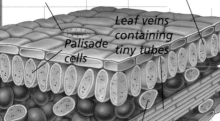

Waterproof wax coat

Upper layer of leaf

Leaf veins containing tiny tubes

Palisade cells

Leaf pores (stomata)

Spongy cells

Lower layer of leaf

◀ *All leaves have veins that carry water.*

▶ *Canadian pondweed oxygenates in water.*

Blowing bubbles

The oxygen made when sunlight breaks up water forms a gas inside the leaf. In plants that grow in the air, the oxygen simply mixes with other air gases and passes out through holes in the leaf without being seen. In water-plants, oxygen forms bubbles of gas that escape from the leaves when the plant makes food (photosynthesizes).

Read further > seaweed
pg9 (k30)

Finding light

Unlike other living things, plants cannot move about to find sunlight but they can grow toward it. Sometimes when plants sprout from their seeds, they are surrounded by other plants that keep them in the shade. Fortunately, the plant stems have tips that are sensitive to light, and they help the plant grow in the right direction.

▲ *While a poppy is growing toward sunlight, the plant uses food that is stored in its bulb.*

Deadly meat-eaters

Carnivorous plants such as this pitcher plant feed on living creatures, including insects and small mammals. The plant attracts the prey with a smell of rotting meat. Once inside the plant's vase-shaped leaves, called pitchers, the victim is dissolved by chemicals known as enzymes.

▶ *The pitcher plant digests and absorbs the bodies of its prey.*

Read further > shady plants / bulbs
pg9 (b28); pg29 (i32; h22)

Read further > digested remains
pg32 (d2)

Most of the world's staple foods, like rice and potatoes, contain starches made by plants

Flowering plants

THE FLOWERING plant group contains more than 250,000 different species, including flowers, herbs, grasses, vegetables and trees (but not conifers, which are gymnosperms – they make their seeds in cones). Flowering plants are divided into two main groups: monocotyledons, which have one cotyledon (food store), such as grasses, rushes, lilies and orchids and dicotyledons, which have two cotyledons – most flowers are formed like this. A plant that lives for one year (see pg24 [m13]) produces a flower at the end of its life. Most plants that live for many years grow flowers once a year. The flower is the part in which seeds are made. Inside each seed is a new plant.

A plant that lives for one year (see pg24 [m13])

● IT'S A FACT

• The Puya raimondii plant lives for 150 years, then grows its first flower and dies.

• Some flowers that grow in the Arctic turn and face the sun all day to absorb as much light as possible and also to keep warm.

• Bougainvillea (a tropical plant) does not have colorful flowers but colorful leaves.

● Inside a flower

Flowers come in different shapes and sizes but most have the same parts. The young flower develops inside a bud, protected by green, leaf-like sepals. As it opens, it displays its outermost parts, the ring of petals, which are usually large and colorful to attract insects. Within the petals is a ring of male parts called stamens. Each has a long, thin stalk – the filament – topped by a brush- or bag-like anther, which contains the male reproductive cells inside their pollen grains. At the center of the flower are the female parts, known as the carpel. This has a sticky pad, called the stigma, on top of a long stalk, called the style, which widens at its base into an ovary. Inside the ovary are the female reproductive cells in their ovules (eggs).

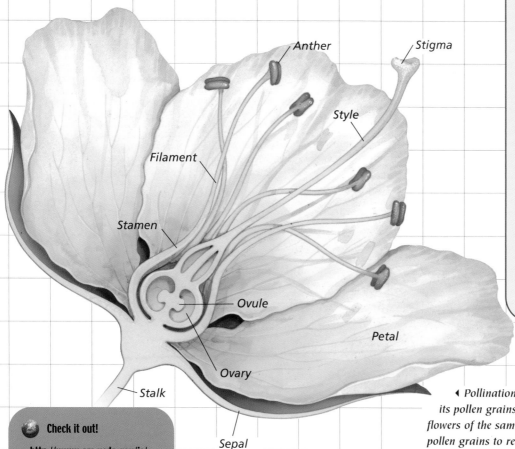

Anther · Stigma · Style · Filament · Stamen · Ovule · Petal · Ovary · Stalk · Sepal

Read further › stamens
pg19 (n30)

◀ Pollination takes place when the male anther releases its pollen grains. These travel to the female stigma on flowers of the same kind. This allows the male cells in the pollen grains to reach the female cells in the ovules.

🌐 Check it out!

• http://www.ars.usda.gov/is/kids/plants/story2/flower.htm

A sunflower is actually a flowerhead made up of many small flowers

1 2 3 4 5 6 7 8 9 10 11 12 13 14 15 16 17 18 19

Catkins

Some trees grow groups of flowers called catkins that hang down from a twig. Stamens hang outside the flowers in a catkin. They make pollen that is blown away by the wind. The wind rocks the catkins and stamens to make the plant release its pollen. The willow and hazel trees grow catkins early in the year before they sprout leaves. The oak grows catkins later in the spring after it has sprouted leaves.

▸ *The alder produces long, slim male catkins, and on the same tree, shorter, rounder female catkins.*

▸▸ Read further › wind pollination
pg19 (n30); pg21 (b32)

GREEN FINGERS

• Many flowers contain both the male reproductive organs (stamens) and the female reproductive organs (ovaries), but some plants have only one or the other.

• Grass flowers do not have petals. The flowers grow together at the top of a stalk.

• The name 'daisy' comes from 'day's eye' since these eye-like flowers open by day and close up at night.

Flowerheads

Some plants grow a large number of small flowers packed tightly together. The small flowers are called florets and they form a disc called a flowerhead. Many plants, such as the daisy and the dandelion, form many flowerheads that fool us into thinking that they are large flowers.

▾ *The wild daisy looks like a single bloom but it actually consists of many flowers. The flowers around the edge have a single petal.*

▸▸ Read further › dandelion
pg21 (b32)

Flower buds

A flower forms inside a bud *(see pg10 [e15])*. Flowers are very delicate when they are growing so the bud has a tough protective covering made from small leaves called sepals. When the flower bursts out of its bud, the sepals bend backwards and may fall off. The sepals of the tomato stay on the plant right through until the fruit is formed. They form the dark, spidery top of a tomato.

◂ *The sepals of the Iceland poppy protect the bud while it is growing.*

▴ *The sepals form the dark, spidery top of a tomato.*

Pollination

BEFORE A flower can make seeds, it must be pollinated. Pollen consists of fine yellow particles that are made by the stamens. The pollen has to be transferred to the stigma of another flower of the same type. The process by which pollen is moved from one flower to another is called pollination. There are several ways in which pollen can travel: it can be carried by insects or other animals, by wind or by water. Flowers that are pollinated by animals have bright colors to attract insects while wind-pollinated flowers are often dull.

IT'S A FACT

• Some people are allergic to pollen. When grass plants release their pollen, these people suffer from a condition called hay fever. This can cause many different reactions such as runny eyes and sneezing.

• Bees collect pollen on their back legs to feed to their young.

Attracting insects

To attract insects, some flowers have brightly colored petals and a powerful scent. Near the base of the petals the flower makes a sugary juice called nectar for insects to drink. As insects search for the nectar, they pick up sticky pollen with their bodies from the stamens or anther *(see pg10 [i17])*. When they visit another flower they transfer the pollen from their bodies to the other plant's sticky stigma.

▶▶ Read further › stamens
pg16 (j14)

◀ *Brightly colored flower petals attract insects, such as this bee, to land on flowers, thus encouraging insects to pollinate them.*

GREEN FINGERS

• A flower can only use pollen that has come from the same type of plant.

• Self-pollination occurs when the pollen moves from the stamen to the stigma in the same flower.

• Cross-pollination occurs when the pollen moves between plants of the same kind.

Rainforest flowers that are pollinated by birds are often red, while those pollinated by moths are yellow or orange

● Bee orchids

Some plants, such as the bee orchid, can pollinate themselves if no insect visits them. The bee orchid's petals are shaped to look and smell like a female bee to attract male bees. But if no bees come along, the orchid can bend over to pollinate itself.

● Spiky pollen

Flowers that are pollinated by animals make a small amount of spiky pollen. The spikes help the pollen stick to the hairs on the bodies of passing insects. The spikes hold the pollen in place as the insect flies between flowers. When the insect brushes against a sticky stigma, the pollen is pulled off.

▼ The bee orchid's stamens bend over and release pollen grains straight onto its own stigma.

▶▶ **Read further › stigma** pg16 (j14)

▶ Spiky pollen sticks to the legs of this Emperor butterfly.

Stamen

Pollen

▶ Magnified pollen grain.

Stigma

lowers higher n the stem re smaller

● Blowing in the wind

Wind-pollinated flowers, such as grass flowers, have no need of bright colors or scents so they are dull and have no smell. The stamens hang out of the flowers so that the wind can blow the pollen away. Later, the flowers put out stigmas to collect any pollen being carried in the air. The catkins (long, dangling flowers) of trees such as the willow and hazel are wind pollinated.

Petals look and smell like a female bee

▶ The lily's stamens are bent like hooks but the stigma and style are straight.

🌐 **Check it out!**

● http://www.cnps.org/kidstuff/ pollin.htm

▶▶ **Read further › catkins** pg17 (b22)

A ragweed flower can release 1 billion pollen grains in a day

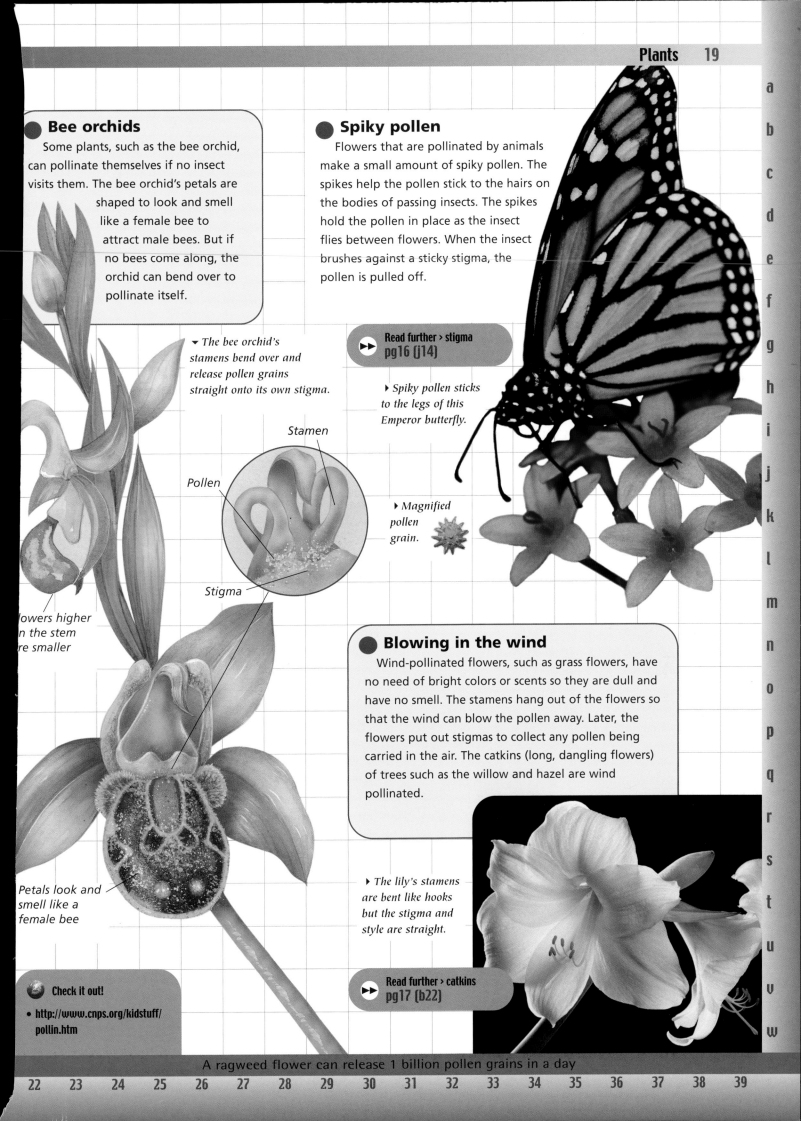

Seeds and fruits

WHEN A pollen grain lands on a stigma, it grows a long, thin tube. The tube grows down through the style and into the ovary. Inside the ovary are one or more egg-shaped ovules. The pollen tube seeks out an ovule to fertilize. After fertilization, the ovule turns into a seed and the ovary turns into a fruit. While these changes are taking place together, parts of the flower, such as the petals and stamens, fall away. The fruit may swell up and become brightly colored and juicy so it is eaten, or it may dry up and form parachutes or wings to enable it to blow away and grow into a new plant.

▼ The seeds of the orange fruit are protected by the swollen flesh.

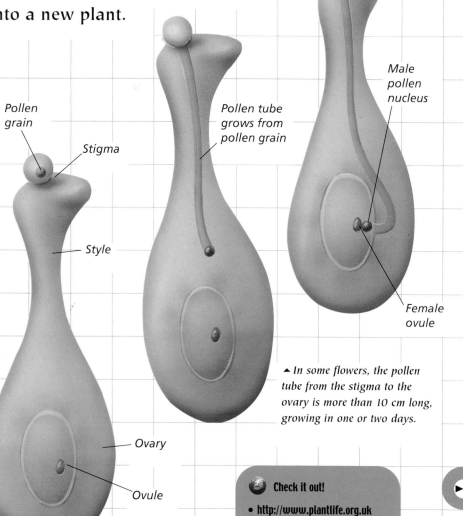

Pollen grain

Stigma

Style

Ovary

Ovule

Pollen tube grows from pollen grain

Male pollen nucleus

Female ovule

▲ In some flowers, the pollen tube from the stigma to the ovary is more than 10 cm long, growing in one or two days.

Making a seed

The center of a pollen grain is called the pollen nucleus. The male pollen nucleus moves down the pollen tube into the ovule. When the pollen nucleus meets the female ovule nucleus, they join together in a process called fertilization. Seeds and fruit form after fertilization has taken place. Food made in the leaves travels along tubes inside the plant until it reaches the seeds and fruit. The food is used to make the parts of the seed and to make the fruit grow. Seeds are kept in a part of the plant called a fruit, which is often swollen and fleshy to protect the seeds from being eaten.

Check it out!
• http://www.plantlife.org.uk

Read further › pollination pg18 (d2)

● One to one

A pollen grain can only fertilize one ovule, so if an ovary has six ovules it will need six pollen grains. A pea pod forms from the ovary of a pea plant. Only fertilized pea ovules can grow as seeds. If ovules are left in the pod, they are not fertilized.

▲ *When the pod is opened, it may contain some seeds that have failed to grow. These are ovules that have not been fertilized.*

● Wind power

Some plants rely on the wind to help move their seeds. The dandelion makes a fruit with a cap of small hairs, which acts like a parachute and keeps the seeds from falling quickly. The sycamore fruits each have a wing and they join together to spin round like helicopter blades as they blow away.

▲ *The sycamore fruit has winged seeds so it can twirl on the wind.*

▸ *Dandelion seeds are found inside fruit attached to hairs that catch the wind.*

Read further › wind pollination
pg19 (n30)

● Berries, pomes and drupes

Fruit such as grapefruit have fleshy layers and are called succulent fruit. Succulent fruit with one seed contained in a hard case, such as cherries and plums, are known as drupes. Succulent fruit containing more than one seed, such as oranges, are called berries. However, raspberries and blackberries are aggregate fruit – they form from many ovaries in just one flower. Each aggregate fruit is made from fleshy beads or drupelets, each containing a single seed. Apples are pomes: they have a fleshy layer containing seeds in a capsule.

● Help from animals

Seeds are carried away from the parent plant so that they do not compete for the same space, light and water. Some fruits make themselves attractive to animals so that they will be eaten. When an animal eats the seeds in the sweet, juicy fruit, they are not completely digested. Instead, they leave the animal's body in the droppings so they have a supply of manure to help the new plants grow!

▾ *Though squirrels eat many kinds of fruit and seeds, which pass out in their droppings, they also store away seeds in places where new plants can sprout.*

● GREEN FINGERS

• The lupin fruit explodes to disperse its seeds.

• The fruits of the goose grass have hooks, which stick onto the fur and feathers of passing animals. The seeds can be carried for many kilometers before they fall to the ground.

• A cereal grain is a dry fruit, which is filled with one seed – grass.

Germination

INSIDE A seed is a tiny plant or embryo waiting to grow. The seed also contains a store of food to help the plant grow. When the seed leaves the parent plant, it is usually dry. This protects the seed from mold and keeps it light in weight. A lightweight seed is able to travel further and land on ground where there is room for it to grow. When the seed settles in the soil, it takes in water, swells up and breaks open so that the new plant can grow out. This process of sprouting is called germination.

● Speeding up sprouting

When tiny plants start to grow, they change stored food in the seeds into substances that build their root, stem and leaves. The speed at which these changes take place is determined by heat. If the seeds are kept in warm conditions such as a greenhouse, the changes take place quickly and the seeds soon sprout. Seeds in cool conditions take longer to sprout. Some seeds, such as those of the ironbark tree, need to be scorched by wildfire before they will germinate.

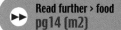

►► Read further › food pg14 [m2]

▸ Gardeners try to mix seeds that flower at different times so the garden is always full of color.

● Seed needs

Not all seeds germinate in the same way. Coconut palm trees grow on tropical beaches and the coconuts fall and germinate on the beach. They can also fall into the sea, sometimes travelling for 2000 km in different ocean currents before washing up on another warm beach, where they may sprout and grow into an coconut palm.

▸ Coconuts are the fruit of the coconut palm tree. The seed has a hollow center that contains a sweet-tasting milk. The seed is contained in a brown, woody shell called a husk.

a
b
c
d
e
f
g
h
i
j
k
l
m
n
o
p
q
r
s
t
u
v
w

IT'S A FACT

• A seed does not always sprout as soon as it forms. It may spend some time before germination, being inactive or dormant.

• The young plant that grows out of a seed is called a seedling.

• The first parts of a seedling that grow are the tips of the shoot and root.

Stages in sprouting

An embryo inside a seed contains a food store (cotyledon), which is protected by the tough, outer casing (testa). The cotyledon then moves to the tiny plant and nourishes it, making the plant grow. As the plant increases in size, the seed case breaks open. The root is the first part to appear and it grows downwards. Later, the tiny shoots appear and grow upwards. In the bean seed shown here the food store stays underground and the shoot makes new leaves above ground.

Read further > cotyledons pg16 (d2)

Keeping growing

When a new plant has used up the food store in its seed, it must find another source of food. The root obtains water and minerals, growing strong enough to hold the plant in place. The shoot grows towards the light so that the first leaves can make food.

▸ *Plants can even get all their needs – sunlight and water – in a crack in a pavement as rainwater carrying minerals seeps in.*

Read further > sunlight pg15 (m22)

GREEN FINGERS

• In the sunflower seed, the food store rises up through the soil on the tip of the shoot to become the first pair of leaves.

• If you plant a seed upside down, the root will still grow downwards and the shoot will still grow upwards. They always grow from the seed in the right direction.

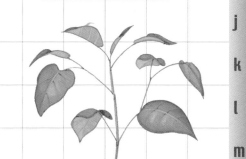

Testa (outer casing)

Cotyledon (food store)

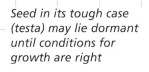

Seed in its tough case (testa) may lie dormant until conditions for growth are right

Seed sends a root down and a shoot up

Shoot opens its first leaves

Stem and roots grow longer, and plant soon begins to grow new leaves

Cycles of life

EVERY PLANT has a life cycle. It begins when the seed first germinates, then it grows and forms flowers that are pollinated, which in turn produce their own seeds. Then the cycle begins all over again. Some plants complete their life cycle in a few weeks – ephemeral plants; others repeat their cycles for hundreds of years – perennial plants. Plants that complete their cycle in one year are annuals. Those that complete their cycle in two years are biennial.

▸ *Annual plants complete their life cycles, from seed to flower to seed again, in one year. In temperate (dry) lands, most of the growth occurs during spring, with pollination in early summer and seed production and growth during spring.*

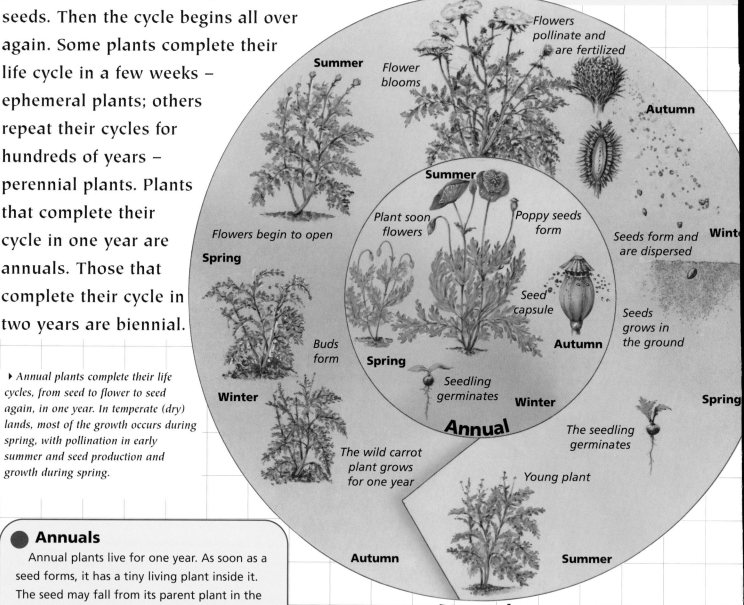

Summer — Flower blooms

Flowers pollinate and are fertilized

Autumn

Flowers begin to open — Spring

Summer — Plant soon flowers — Poppy seeds form

Seeds form and are dispersed — Winter

Seed capsule

Seeds grows in the ground

Buds form — Spring

Autumn

Winter

Annual

Winter

Seedling germinates

The seedling germinates — Spring

The wild carrot plant grows for one year

Young plant

Autumn — Summer

Biennial

● **Annuals**

Annual plants live for one year. As soon as a seed forms, it has a tiny living plant inside it. The seed may fall from its parent plant in the autumn and lie all winter in the ground. The following spring the tiny plant bursts out of the seed and grows steadily for a few months. Finally it makes flowers, and after its seeds have formed and spread, the plant dies.

▸▸ **Read further › seeds pg23 (b22)**

▲ *Biennial plants take two years to grow from seed to a mature plant. The part-grown plant must be able to survive the harsh conditions of the cold or dry season, then continue its development the year after to flower and make seeds.*

Ephemerals

The seed of the shepherd's purse plant sprouts soon after it lands in the soil. Such plants that grows quickly, flower, set seeds and die in a few weeks are called ephemeral plants. There are many life cycles of the shepherd's purse, one after the other throughout the year. This means that shepherd's purse seeds are being spread around at regular intervals so there is a better chance that some may find a place to grow.

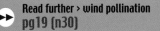
Read further › wind pollination pg19 (n30)

▲ *Daffodils are perennials – they have bulbs and bloom every year.*

Perennials

When a perennial plant has finished flowering, it may lose some of its parts that are above ground (the flower) but the roots live on underground during the winter. The new shoots sprout the following spring. Perennials, such as tulips and daffodils, have a thick underground food store, called a bulb, that stays alive during the winter when the rest of the plant has died away.

Read further › exploding flowers pg31 (o30)

Biennials

Biennials have life cycles that take two years to complete. In the first year, or growth season, the seed sprouts and the plant grows. During this time the leaves make plenty of food, which is stored in the plant. In the second growth season biennial plants use the stored food to help them grow again, bloom with flowers and produce seeds before dying.

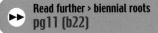
Read further › biennial roots pg11 (b22)

▲ *In the first year of their life, beetroots develop leaves and a fleshy, red root – they are harvested before the next year.*

GREEN FINGERS

• Ephemerals (short-lived plants), such as chickweed and groundsel, can take over a bare patch of soil quickly and become weeds (plants growing in the wrong place).

• A potato plant grows underground stems, which swell up with stored food to form tubers, called potatoes. Each stem can grow into another plant the following year.

The spider plant grows tiny plants on stalks that branch among its leaves

Woody plants

T HE ROOTS and shoots of most plants are made of soft materials, but some plants make their roots and shoots from a very hard material—wood. These woody plants are trees and bushes. Unlike other plants that die at the end of the growing season, wood is so strong that it allows the plant to keep on growing and become larger and larger every year. The woody stems are covered in bark to protect the tree from animals that might try to eat it. Bark also protects the tree from the bitterly cold winter weather by making a warm insulating layer over the surface.

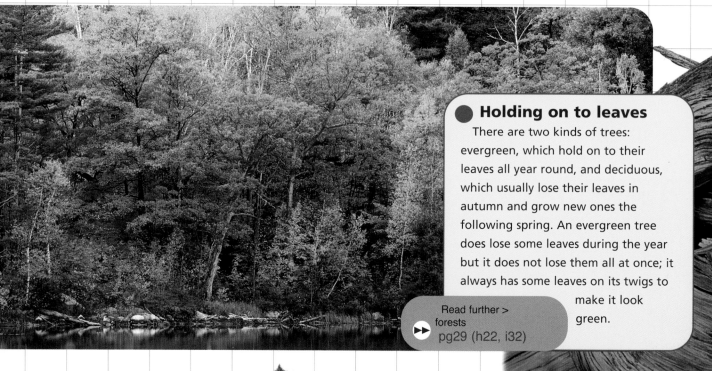

● Holding on to leaves

There are two kinds of trees: evergreen, which hold on to their leaves all year round, and deciduous, which usually lose their leaves in autumn and grow new ones the following spring. An evergreen tree does lose some leaves during the year but it does not lose them all at once; it always has some leaves on its twigs to make it look green.

Read further >
▶▶ forests
pg29 (h22, i32)

To scale

This coast redwood is the fourth highest tree ever measured – it is about the length of a soccer field

Each square = 25 m

Check it out!
• http://www.treecouncil.org.uk

Coast redwood is about 112 m high

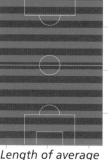
Length of average soccer field is 92 m

Some pine needles can grow to a length of 30 cm

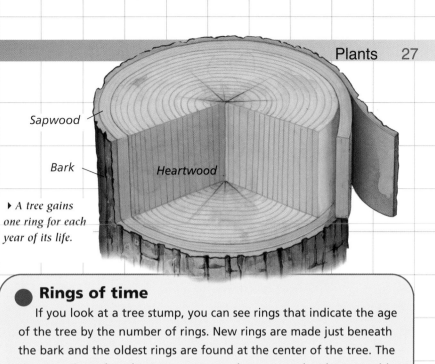

Sapwood

Bark

Heartwood

▶ *A tree gains one ring for each year of its life.*

Old is normal

Woody plants can live for a long time, so it is normal to see very old trees or bushes. The oak tree does not start making flowers until it is 50 years old and may then live for hundreds of years after that. The bristlecone pine, which grows in North America, is one of the oldest species of tree in the world – almost 5000 years old.

◀ *The bristlecone pine has a long lifespan because it grows very slowly and in cool, dry areas, such as Nevada, Utah and California.*

▶▶ Read further > long lives
pg25 (j34)

Rings of time

If you look at a tree stump, you can see rings that indicate the age of the tree by the number of rings. New rings are made just beneath the bark and the oldest rings are found at the center of the tree. The rings are tiny tubes that carry water to the tree. As they become older, the rings still provide support. Heartwood is the dark, dead wood in the center of the trunk. Sapwood is pale living wood, where tiny pipes called xylem carry sap from the roots to the leaves.

▶▶ Read further > stems and trunks
pg11 (b34)

SUPER GIANTS

Trees	Country	Height
• Eucalyptus	Australia	132.6 m
• Douglas fir	USA	126.5 m
• Mountain ash	Australia	114 m
• Coast redwood	USA	112 m

IT'S A FACT

• Cork is made from the thick, light bark of cork or other trees.

• Broad–leaved trees grow flowers; conifers grow cones.

• The tree with the toughest wood is the ombu tree, which grows in Argentina.

Pine leaves are needle-shaped

Leaves act like fingerprints

Leaves come in many different shapes and sizes, but there are two main types: needle-shaped leaves and broad leaves. Conifers, such as the pine, make needle-shaped leaves; broad leaves are made by trees and other woody plants in the flowering plant group. A broad leaf may be just a single leaf, such as an oak tree leaf, or be made from a group of leaflets, such as the ash. You can identify a particular type of tree by the shape of its leaf.

Oak leaves are lobe-shaped

▶▶ Read further > pine cones
pg9 (g22)

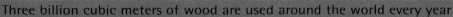

Forest life

WHEN TREES grow together in large groups they form forests. There are three main types of forest: coniferous, temperate and tropical rainforests with broad-leaved trees. Each kind of forest grows in different places around the world, depending on weather conditions. When trees grow close together, such as in a rainforest, the leaves on the branches stop most of the light reaching the forest floor. Plants that live on the ground have adapted to surviving in the shade beneath the trees.

Emergent layer

Canopy

● Tropical rainforests

Rainforests grow in the hot, wet weather conditions around the Equator – the nearest part of the Earth to the Sun. The trees in a rainforest are mostly broad-leaved, evergreen trees. Most of the trees grow up to 30 m above the ground and their branches lock together to form a canopy. Many small plants, called epiphytic plants, such as orchids, ferns and bromeliads, live on the branches of rainforest trees. They have short roots to hold them in place and gather the water they need in cups made from their leaves.

Understorey

Undergrowth

▶▶ Read further > roots pg11 (b22)

Forest floor

A single large coniferous tree may have 20 million needle-like leaves

1 2 3 4 5 6 7 8 9 10 11 12 13 14 15 16 17 18 19

▼ *Coniferous forests grow for hundreds of kilometers in northern parts of North America, Europe and Asia, usually high up on mountainsides.*

Temperate forests

Temperate forests grow where summers are warm and winters are cool. They are found in parts of North America, Europe, China, Australia and Japan. Trees in temperate forests lose their leaves in winter. Therefore, in spring, before new leaves grow, light reaches the forest floor giving flowering plants a chance to flower.

Read further > sunlight
►► pg15 (m22)

▲ *Bluebells bloom on the forest floor before the trees grow new leaves in spring.*

Coniferous forests

Coniferous forests grow where the summers are short and warm but the winters are long and cold. As the trees grow very close and they keep their leaves all year round, very little light reaches the forest floor so only small plants such as mosses can grow. The soil in coniferous forests usually freezes once a year, making it difficult for plants to obtain water.

Read further >
mosses
►► pg9 (b28)

GREEN FINGERS

• One 23-hectare (about 57 acres) area of Malaysian rainforest has 375 species of tree with trunks thicker than 91 cm.

• The floor of a coniferous forest is covered in a thick layer of dead leaves because they take a long time to rot in the cold, dark conditions.

• Rainforest trees are covered with epiphytes – plants whose roots never reach the soil but take water from the air.

Check it out!

• http://www.woodland-trust.org.uk/

Can forests survive?

Forests provide people with large amounts of wood. Every second, an area of rainforest the size of a soccer field is cut down (logged) for timber. Within the next 100 years all the rainforests could be destroyed. Planting more trees and recycling paper materials are two ways that can help rainforests to survive.

▸ *Logging can destroy whole hillsides in as little as one day.*

Read further > tree growth
►► pg27 (f29)

In Australia there are over 400 kinds of eucalyptus trees

a b c d e f g h i j k l m n o p q r s t u v w

Grasslands and desert

GRASSLAND FORMS where there is not enough rain for trees to grow. There is usually a long, dry season and a short season, when it rains or even snows. In the dry season, plants and seeds may lie dormant (asleep), but when rain falls the grassland briefly blooms with green leaves and brightly colored flowers. In hot countries, during the dry season, plants face the threat of fire, but the fire does not destroy the roots so the grass can grow back again. A desert is a place where very little rain falls: there may be nine months without rain, then three months with a tiny amount. Deserts can be very hot in the daytime and very cold at night so it is difficult for plants to thrive without the mineral-rich soil and water that they need to survive.

IT'S A FACT

• The continent with the largest area of desert is Asia.

• The world's biggest desert is the Sahara in north Africa. It is 5000 km across at its widest point and up to 2250 km from north to south.

The savannah

The savannah is a huge area of grassland that covers most of Africa. It covers a much larger area than the rainforest or the deserts. The tallest grass on the savannah is elephant grass, which reaches 3 m in height. Some trees, such as the acacia and the baobab, which can store water and have fire-resistant trunks, grow in small numbers on the grassy plains.

Read further > rainforests pg28 (o29)

▼ *Acacia trees provide a small amount of protection for herds of zebra in the blazing sun and drying winds of the African savannah.*

There are 8000 different kinds of grass plant

1   2   3   4   5   6   7   8   9   10   11   12   13   14   15   16   17   18   19

Prairies contain a wide variety of grass including cottonwood, switch grass, needle grass, panicgrass, asters, Idaho fescue and big bluestem grass.

● Prairies and steppes

Large areas of temperate grassland in North America are called prairies – in Russia they are called steppes. Here there is not enough rain all year round to allow trees to grow but hundreds of grasses, crops and shrubs thrive.

▶▶ Read further > crops
pg34 (h15)

GREEN FINGERS

• Joshua trees, native to North American deserts, can grow to about 9 m high.

• A large cactus can take in 1 ton of water in a day after it has rained.

• The biggest cactus is the saguaro of southern Arizona, southeast California and northwest Mexico. It can grow up to 20 m tall and 1 m thick.

● Plants like pebbles

Pebble plants grow in deserts in southern Africa. They develop thick, round leaves, colored like stones and pebbles, that camouflage and protect them against animals that might eat them.

▶▶ Read further > animals
pg21 (k31)

● Cacti

The largest plants in most deserts are succulents. Cacti are a particular type of succulent that grow mostly in North and South American deserts. Down the sides of the cactus are grooves that fill out when it takes in water. Cacti do not have broad leaves to make food because they would lose too much water; instead they use their thick, green, fleshy stems as leaves to store water and convert sunlight into food. A waxy layer on the stem reduces water evaporation.

▶▶ Read further > path of water
pg12 (m2)

▲ *Needle-like spikes on the cactus stem stop animals from biting their way into the plant to reach water.*

● Exploding seedheads

Some plants cannot store water so they die when the desert dries up. Their seedheads, packed with seeds, stand on dry stalks until the next rain. When the water soaks into the flowerhead it makes parts twist and turn so that the flowerhead explodes and disperses its seeds. The seeds take in water and sprout quickly to make new plants.

▶▶ Read further > seed dispersal
pg21 (b32)

▶ *After rain, cactus flowers bloom briefly, adding splashes of colour to the desert.*

The quiver tree breaks off its branches to save water

Fungi and lichens

FUNGI ARE simple, plant-like organisms that grow up from the ground. There are more than 50,000 different types of fungi, including mushrooms and toadstools, yeast and molds. Fungi are not plants because they have no chlorophyll with which to make food from the sun, and no leaves or flowers. Fungi feed either from other plants and animals or from dead matter that has rotted in the soil. Lichens, which grow on rocks and tree bark, are a combination of fungi and algae, which lack proper stems, roots, leaves and flowers.

● IT'S A FACT

• The gills of a mushroom can produce 16 billion spores.

• Molds are fungi that grow on food such as bread and fruit.

• A truffle is an edible fungus that grows underground.

● Fairy ring

As a fungus grows, its hyphae (tiny threads that grow through the soil) form a tangled mass of fibers called a mycelium. This grows in all directions through the soil. In the fairy ring mushroom, the mycelium grows out in a disc-shape in the soil. The fruiting bodies grow up along its edge and make a ring of mushrooms.

● The fruiting body

The part of the fungus that grows above ground is called the fruiting body. A mushroom or toadstool is divided into two parts: the stalk and the cap. When the cap is fully formed it spreads out like an open umbrella. Under the cap are rows of vertical sheets called gills that are used for reproduction. They release microscopic spores that grow into new fungi.

▶▶ Read further › spores pg9 (b28)

▶▶ Read further › growing taller pg15 (m22)

▶ *As the fairy ring mushroom's hyphae grow outwards, they have a naturally fertilizing effect on the soil because they break down decaying matter and make nutrients available for themselves.*

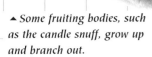

▲ *Some fruiting bodies, such as the candle snuff, grow up and branch out.*

● Working together

Lichens survive in harsh conditions because they are made from both fungi that absorb water and algae that produce food.

🌐 Check it out!

• http://www.herb.lsa.umich.edu/kidpage.
• http://www.mykoweb.com

Mold and mildews are relatives of fungi

Earth star

When an earth star first grows above ground, it looks like an onion. As it grows, its outer wall splits into six or more parts that bend backwards on the ground, making the fungus look like a star. At the center of the fungus is a ball-shaped object with a hole in the top. When the fungus is fully grown, spores escape from the hole into the air.

▲ *The earth star grows in soil or rotting wood.*

Read further > opening up
pg17 (r27)

Puffball

Puffballs grow above ground in woodlands in late summer and early autumn. They are greyish-white at first but as they grow to full size they become yellowish-brown. Inside the puffball are millions of spores. When they are ready to leave, holes develop on the surface of the puffball. As the puffball is blown by the wind or hit by raindrops it is squeezed slightly and the spores escape through holes into the air. Huge numbers of spores are released simultaneously.

Read further > woodland
pg29 (h22)

▲ *When they burst, puffballs puff out their spores in all directions.*

Colorful growth

Many fungi are white but some have red, yellow, orange, brown, blue and purple parts. Some even glow in the dark! The color does not help in telling you whether the fungus is poisonous or not and for this reason wild fungi should be avoided. For example, the field mushroom and destroying angel are both white, yet the field mushroom is edible and the destroying angel is deadly poisonous. The red cap of the fly agaric warns that it is poisonous while the honey mushroom can be cooked and eaten.

Fly agaric

Honey mushroom

Destroying angel

Field mushroom

Read further > color
pg11 (n22)

Life cycle of a fungus

Thread-like hyphae grow from spores into the soil. The tips of some hyphae swell or enlarge, then the swollen tip grows above ground as the fruiting body. Finally, the cap opens and releases spores before dying.

Cap releases spores

Spores germinate

Fruiting body

Mycelium grow from spores

GREEN FINGERS

• Types of algae include bladderwrack, kelp (seaweed), oarweed and carragheen.

• Some kinds of mold are used to give certain cheeses, such as Danish Blue, their taste.

• The penicillium fungus is a mould that makes a chemical to kill bacteria that grow around it. The chemical is used to make the antibiotic drug called penicillin.

One of the most deadly fungi is the death cap

People and plants

PLANTS ARE vitally important to life on our planet because they make oxygen that all animals – including humans – need to breathe. Plants also provide food that is eaten by many animals and people. The first people gathered wild plants from the countryside around them for food. Today we grow plants in fields and plantations to make harvesting easier. Plants also provide us with useful materials such as wood, cotton, paints, rubber and medicines. Plant leaves, stems, flowers and sap have been used since the earliest times for medicinal purposes and, even today, plant-based remedies are popular across the world.

IT'S A FACT

- The pain-relieving drug morphine is made from the opium poppy.

- Linen is made from fibers in the stem of the flax plant.

Growing crops

On a farm the growing of a crop begins by preparing the soil before sowing the seeds. When the seeds have sprouted, the young plants may be given fertilizer to help them grow and may be sprayed with pesticides to prevent insects and fungi from attacking them. When the crop is fully-grown, it is harvested.

5. Seeds sprout before winter sets in, but they do not begin to ripen until the following spring

1. Wheat is harvested (gathered) in late summer

2. A few days after harvesting, the soil is cultivated (prepared) to get rid of unwanted weeds

3. After cultivation, soil is plowed and harrowed (broken up)

4. About six weeks after harvesting, seeds are sown in prepared soil

Read further › fungi pg32 [d2]

Check it out!
- http://www.edenproject.com

It takes 170,000 saffron crocus flowers to make 1 kg of the spice called saffron

1 2 3 4 5 6 7 8 9 10 11 12 13 14 15 16 17 18 19

Rice growing

Since about 3000 BCE rice has been cultivated, and it is now the basic food for about half the world's population. Rice seeds are first sown in drier soil, but when the young plants (seedlings) are about two months old they are planted in paddy fields that are flooded with up to 10 cm of water.

▶▶ **Read further > underwater seeds**
pg9 (k30); pg15 (b34)

▲ *Paddy fields form level strips called terraces along hillsides so the amount of water in them can be controlled.*

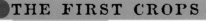

THE FIRST CROPS

Crop	First cultivated
• Pea	9000 BCE
• Wheat	7000 BCE
• Rye	6500 BCE
• Bean	5000 BCE
• Barley	4500 BCE

▼ *Many of the clothes we wear are made using fibers from the cotton plant.*

Shelter and clothes

Wood was probably the first material from a plant to be used. People used it to make shelters and spears. Then about 5000 years ago people began weaving cotton – a fiber that grows on the seeds of the cotton plant – to make clothes. The fibers, which form a fluffy coating around a seed, are spun into yarn before cloth can be made.

◀ *The bolls picked for cotton develop from the seed pod left when the petals of the cotton flower drop off in summer.*

▶▶ **Read further > wood**
pg29 (o29)

◀ *The bitter-tasting century plant was once used to help people with fevers and digestion problems.*

Medicinal purposes

For almost 50,000 years people have used plants to treat illness. Aspirin, a widely used painkiller, was first made from willow bark. Today, scientists work with people who live in tropical rainforests to discover more plants that can help to prevent diseases and cure ailments.

▶▶ **Read further > willow**
pg17 (b22)

Burning rubber

Rubber is derived from the sap of the rubber tree. The sap oozes from a slit in the bark into a collecting cup. In its natural form, rubber (latex) is very soft and flexible. Chemicals are used to treat it in a process called vulcanizing, making the substance harder and tougher for making, for example, rubber tires.

Rubber tree

Sapota trees in Central America produce a gum used to make chewing gum

a b c d e f g h i j k l m n o p q r s t u v w

Glossary

Allergic A condition some people experience when they come into contact with a substance such as pollen.

Annual A flowering plant that completes its life cycle in one year.

Bark A corky substance that grows over the trunk, branches and twigs of trees. It keeps the inside of the tree warm.

Biennial A flowering plant that takes two years to complete its life cycle.

Broad-leaved tree A tree that grows leaves that are wide and flat. It may be an evergreen or deciduous tree.

Bud A lump on a stem containing a tiny stem, leaf or flower that is ready to grow.

Cell A tiny part of a plant's body. The body is made from large numbers of cells. The cells have different shapes to perform different tasks to keep the plant alive.

Cellulose A substance made by plants that gives support to all parts of it.

Chlorophyll A green substance found in plants, which traps sunlight that falls onto the leaves. This helps the plant make food.

Compost A type of soil that is rich in decayed plant material, which can supply essential minerals to growing plants.

Conifer A tree, usually evergreen, that produces cones to make pollen and seeds.

Cotyledon The food store that is contained in every new seedling.

Deciduous tree A tree that loses all its leaves at one time, spends some time with bare branches, then grows new leaves.

Dicotyledon A plant that has two cotyledons, for example a pea.

Ephemeral A plant with a life cycle that only takes a few weeks to complete.

Evaporate A process by which a liquid changes into a gas at normal temperatures without boiling.

Evergreen tree A tree that does not shed its leaves all at once and so remains green throughout the year.

Fertilization The process by which the nucleus of the pollen grain fuses with the nucleus of the ovule to make a seed.

Fruit The part of the plant that develops from the ovary in the flower after fertilization has taken place.

Fungi Living things, most of which have tiny hairs, that feed on dead plants and animals and decompose them.

Germination The process that occurs when the seedling breaks out of its case, sending out its root and shoot.

Insect An animal with six legs and usually two pairs of wings.

Leaf A part of the plant in which food is made. It may be wide or narrow.

Minerals Substances in the soil that a plant needs for healthy growth.

Monocotyledon A plant with only one cotyledon, for example a grass plant.

Perennial A plant that lives for many years.

Photosynthesis The process by which plants make food using water, sunlight and carbon dioxide from the air.

Pollen Tiny capsules made by the stamens. To reproduce, plants must pass pollen between flowers. Pollen is carried by the wind or other insects to other plants.

Pollination The process by which the pollen moves from the stamens to the stigma of the same flower or a different flower on the same kind of plant.

Protein foods Foods, which contain substances for making the different parts of a plant such as the root or leaves.

Root The part of the plant that holds the plant in the soil and takes up water and minerals for the plant to use.

Sap The liquid that moves through the root and stem of a plant.

Seed A capsule, which forms from the ovule after fertilization and contains a tiny plant with a foodstore.

Sepal A small leaf that forms part of the covering of the flower bud.

Spines Long, thin pointed parts of a plant that defend it from plant-eating animals.

Spore A tiny capsule made by non-flowering plants such as mosses and fungi. It contains a piece of the parent plant or fungus, which can break out and grow into a new plant.

Stamen The part in the flower that produces pollen.

Stem The part of the plant that supports the leaves and flowers and carries water and minerals. It also carries food made in the leaves to other parts of the plant.

Stigma The part of the flower on top of the ovary, which receives pollen.

Stomata Holes on the underside of a leaf that let air and water in and out.

Temperate forest A forest that grows in a region with warm summers and cool winters.

Testa The protective outer casing of a seed.

Tropical rainforest A forest that grows where the weather is always hot and wet.

Wax A transparent waterproof coating found on many plants, especially leaves.

Wood A material made by trees that contains large numbers of tiny fibers.

Index

Entries in bold refer to main subjects; entries in italics refer to illustrations.

The publishers would like to thank the
following artists who have contributed to this book:
Ron Hayward, Roger Kent, Alan Male, Janos Marffy
Mike Saunders, Guy Smith, Rudi Vizi

All other photographs are from
Corbis, Corel, digitalSTOCK, John Foxx, MKP Archives, PhotoAlto, PhotoDisc

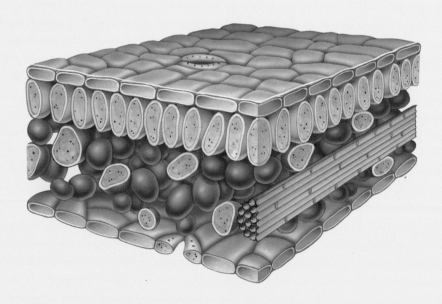